Andréa Bicca Noguez Martins
André Pich Brunes
Dario Munt Moraes

Aspects of seed storage

Andréa Bicca Noguez Martins
André Pich Brunes
Dario Munt Moraes

Aspects of seed storage

Preserving physiological quality

ScienciaScripts

This book is a translation from the original published under ISBN 978-3-330-76997-7.

Publisher:
Sciencia Scripts
is a trademark of
Dodo Books Indian Ocean Ltd. and OmniScriptum S.R.L publishing group

120 High Road, East Finchley, London, N2 9ED, United Kingdom
Str. Armeneasca 28/1, office 1, Chisinau MD-2012, Republic of Moldova, Europe
Printed at: see last page
ISBN: 978-620-8-13485-3

SUMMARY

1 <u>Introduction</u>

Andrea Bicca Noguez Martins

Andre Pich Brunes

Dario Munt de Moraes

According to studies carried out by various researchers, the annual damage that the world economy suffers as a result of post-harvest losses is very large. The most frequent cause of storage losses is attacks by insects, fungi and rodents. There are also losses of intrinsic qualities such as appearance and taste in the case of beans for consumption and, when it comes to seeds, their ability to germinate and produce a vigorous and healthy plant.

The purpose of storage is to preserve the seeds, preserving their physical, physiological and health qualities, for subsequent sowing and obtaining healthy plants after germination. The purpose of stored seeds can be diverse, ranging from the formation of commercial plantations to gene banks for native forests. Depending on the purpose, it may be necessary to store them for short or long periods. The behavior of seeds during storage is a function of the factors that affect their preservation, such as temperature, relative humidity, the degree of seed moisture and the type of packaging used.

The main means used for storing seeds are the cold chamber, the dry chamber and the cold and dry chamber, which adapt to most conditions. During storage, seeds breathe continuously, consuming their reserves and transforming them into water, heat and carbon dioxide. The loss of these chemical compounds during storage must be reduced to a minimum by implementing seed handling processes that ensure the quality of the stored product, since proper seed storage avoids qualitative and quantitative losses.

Seed storage has been the subject of studies, especially in humid and hot regions, where these problems are aggravated. In order to help producers who need to store seeds, we have published this bibliographical review on seed storage.

2 Seed storage

Andrea Bicca Noguez Martins

Andre Pich Brunes

Dario Munt de Moraes

Seed storage became a primary activity when humans stopped being nomadic and started growing their own food, needing to preserve seeds for the next planting. This initially involved protecting them from birds, insects and micro-organisms, and then aspects related to germination and the environmental factors that influence longevity (MEDEIROS & EIRA, 2006).

The purpose of storage is to preserve seeds, preserving their physical, physiological and health qualities, so that they can be sown later and healthy plants can be obtained after germination. Depending on the objective, it may be necessary to preserve them for short or long periods (BENEDITO, 2010). The fundamental environmental conditions for maintaining seed viability during storage are humidity and temperature. The length of storage will depend on how the seeds are to be used in the future, with a short period of up to six months, a medium period of up to five years and a long period of more than five years being considered (FOWLER; MARTINS, 2001).

Storage conditions are decisive in guaranteeing the physiological quality of seeds and, although their quality cannot be improved, good conditions during this period will help to keep seeds viable for longer, slowing down the deterioration process (ALMEIDA et al., 2010). This has led seed producers to worry about using techniques that can minimize the factors that cause deterioration (LIMA et al., 2014).

Seed deterioration begins after physiological maturity and the biggest challenge is to ensure that seeds still retain their quality after a certain period. Therefore, the aim is to maintain seed quality during the storage period (VILLELA & PERES, 2004).

Modern preservation techniques can only prolong the useful life of the seed during storage. However, the deterioration process will be accelerated when the stored seed is of low initial quality, which is explained by the fact that seeds belong to the category of deteriorable but not perishable products (ALMEIDA et al., 2010).

The ability of a seed to maintain its quality during storage depends on the inherent longevity of the species, its initial quality and the environmental storage conditions (CARVALHO & VILLELA, 2006). Therefore, seed can be produced under a rigorous inspection system, properly harvested, and processed to the highest purity, but it can be lost if stored under inadequate conditions or with a high water content.

Therefore, low seed water content and low temperature, combined with low relative humidity in the storage environment, are important factors for maintaining quality for a longer period (NOBRE et al., 2013).Several factors influence the preservation of seed vigor and viability during storage, the main ones being: temperature and relative humidity, types of packaging and storage duration (CARNEIRO & AGUIAR, 1993; CARVALHO &NAKAGAWA, 2000).

2.1.   Storage environment

Relative humidity, followed by temperature in the storage environment, are the main factors affecting the physiological quality of seeds, as they are directly related to their metabolic processes (POPINIGIS, 1985). Because they are hygroscopic, the water content of seeds is constantly influenced by the relative humidity of the surrounding air, always seeking hygroscopic equilibrium (HARRINGTON, 1972). The length of time it takes for the moisture content of seeds to come into equilibrium with the relative humidity of the environment (hygroscopic equilibrium point) depends on the species and, above all, the temperature. The higher the temperature, the faster hygroscopic equilibrium will be reached (HARRINGTON, 1972; TOLEDO & MARCOS FILHO, 1977).

The respiration rate of the seed is influenced by its water content, temperature, membrane permeability, oxygen tension and light (POPINIGIS, 1985). Respiration implies the loss of dry mass and gas exchange, and the methods used to measure respiration are based on determining these characteristics. However, measuring the change in dry mass requires a large amount of material, as well as being considered a somewhat time-consuming analysis to obtain the result, given that the plant material must be completely dried in an oven (MARENCO & LOPES, 2007).

The higher the temperature and relative humidity, the higher the respiration rate of the seeds. The consequences of increased respiration are the moistening and heating of the seed mass, which is aggravated when combined with the action of microorganisms and insects (BAUDET &VILELA, 2006). The result is the rapid consumption of reserves, causing loss of mass and a drastic decline in germination and vigor. The drier and colder the storage environment, within certain limits, the greater the chances of prolonging seed preservation (TOLEDO & MARCOS FILHO, 1977).

There is an increase in the respiration rate proportional to the increase in temperature, which is dependent on the water content of the seeds (SILVA, 2008). With water contents above 14% (b.u.), respiration increases rapidly in most cereals, causing them to deteriorate (SMANIOTTO et al, 2014). Reducing the temperature is an economically viable practice for preserving the quality of stored seeds (DEMITO & AFONSO, 2009).

The main environments for storing seeds are cold rooms, dry rooms and cold and dry rooms, which adapt to most conditions (FREITAS; 2009).

Studies carried out on amaranth seeds, evaluating the influence of the storage environment on the physiological quality of the seeds, revealed that storing amaranth in a cold room was more efficient in preserving the physiological quality of the seeds (NOBRE et al., 2013).

2.2. Packaging

Modern technology aims to preserve the original characteristics of seed lots and, to do this, a series of methods and materials are used. The physical (size, color, weight, moisture content, health, physical purity) and physiological (germination, vigor, chemical composition, latency) characteristics, the environment and the storage period influence the choice of these methods and materials for good preservation. The technology used to manufacture the packaging used to store the seed must reduce the speed of the deterioration process and the loss of its physiological quality (POPINIGIS, 1985).

The type of packaging used to store the seeds is of great importance in preserving their viability and vigor (CROCHEMORE, 1993). This quality consists of the sum of attributes that indicate the seed's ability to perform vital functions such as germination, vigor and longevity (PESKE & BARROS, 1998). In this way, physiological potential becomes an important component in quality control programs aimed at guaranteeing satisfactory seed performance.

Depending on the type of packaging used to store the seeds, there may be a greater or lesser exchange of water vapor between the seeds and the atmosphere, which can compromise their vigor (MARCOS FILHO, 2005). For this reason, they deserve special attention in terms of their viability during the marketing period, demonstrating whether the seeds in each batch have the capacity to form normal seedlings, capable of developing and expressing their genetics in the field.

All seed intended for sowing must be carefully processed and preserved during the storage period until it is used, in order to guarantee the preservation of its physiological quality (ALMEIDA et al., 1997). A fundamental factor that is of great value for the establishment of crops is the use of high quality seeds that provide an adequate plant stand in the field, enabling high yields (SILVA et al., 2008).

Packaging can be divided into permeable, semi-permeable and impermeable, depending on the moisture exchange that can occur between the seeds and their environment (BAUDET, 2003). They are important for protecting seeds from attack by insects and other animals, as well as facilitating handling, transportation and making better use of storage space (POPINIGIS, 1985).

When seeds are stored in permeable packaging (paper, jute, cotton and woven plastic), their water content varies according to changes in relative humidity, due to the fact that they are hygroscopic. In semi-permeable packaging (thin plastic or polyethylene bags, 0.075 to 0.125 mm thick, and multiwall paper bags laminated with polyethylene), there is some resistance to change, but nothing that completely prevents moisture from passing through, in impermeable packaging (heat-sealed plastic bags over 0.125 mm thick, aluminum packets and aluminum cans, when well sealed), there is no influence of the humidity of the outside air on the seed (POPINIGIS, 1985).

The choice of packaging must take into account the climatic conditions under which the seeds will be stored until sowing, the method of marketing, availability and the mechanical characteristics of the packaging (CARVALHO & NAKAGAWA, 2000).

LITERATURE CITED

ALMEIDA, F. A. C. *et al.* Study of techniques for the storage of five oilseeds in environmental and cryogenic conditions. **Revista Brasileira de Produtos Agroindustriais**, v. 12, p. 189-202, 2010.

ALMEIDA, F. de A. C.; MATOS, V. R.; CASTRO, J. R. de; DUTRA, A. S. Evaluation of seed quality and conservation at producer level. In: ALMEIDA, F. de A.C.; HARA T. CAVALCANTI MATA, M.E.R.M. **Armazenamento** de **graos e sementes nas propriedades rurais**. Campina Grande: UFPB/ SBEA, 1997, 201p.

BAUDET, L.M.L. Seed storage. In: PESKE, S.T.; ROSENTAL, M.D.; ROTA, G.R. (ed.). **Seeds**: scientific and technological foundations. Pelotas: Ed. Universitaria - UFPel, 2003. p.370-418.

BAUDET, L; VILLELA, F.A. Seed Storage. In: PESKE, S.T.; LUCCA FILHO. O.A.; BARROS, A.C.S.A. (Ed.). **Seeds**: scientific and technological foundations. 2. ed. Pelotas: Ed. Universitaria, 2006, p.427-472.

BENEDITO, C. P. **Storage and viability of catanduva *(Piptadenia moniliformis* Benth) seeds**. 2010. 63f. Dissertation (Master's Degree) - Universidade Federal Rural do Semi-Arido, Mossoro, 2010.

CARNEIRO, J. G. A.; AGUIAR, I. B. Seed storage. In: AGUIAR, I. B.; PINAPINA-RODRIGUES, F. C. M.; FIGLIOLIA, M. B. (Ed.). **Tropical forest seeds**. ABRATES, 1993. p. 333-350.

CARVALHO, N.M.; NAKAGAWA, J. **Seeds**: science, technology and production. 4. ed. Campinas: Fundapao Cargill, 2000. 588p.

CROCHEMORE, M.L. Preservation of blue tremopo seeds in different packages. **Revista Brasileira de Sementes**, Londrina, v.15, n.2, p.227-232, 1993.

DEMITO, A.; AFONSO, A. D. L. Quality of artificially cooled soybean seeds. **Engenharia na Agricultura**, v.17, p.7-14, 2009.

FOWLER, J.A.P.; MARTINS, E.G. **Manejo de sementes de especies florestais**. Colombo: Embrapa Florestas, 2001. 76p. (Documents, 58).

FREITAS, R.A.; NASCIMENTO, W.M. Accelerated aging test on lentil seeds. Revista Brasileira de Sementes, v. 28, n. 3, p.59-63, 2006.

HARRINGTON, J.F. Seed storage and longevity. In: KOZLOWSKI, T.T. **SeeBiology**. New York: Academic Press, 1972. v.3, p.145-245.

LIMA, S. M. P.; GUIMARAES, R. M.; OLIVEIRA, J. A.; VIEIRA, M. G. G. C. Effects of conditioning times and temperatures on the physiological quality of coffee seeds (Coffea arabica, L.) under ideal heat stress conditions. Ciencia Agrotecnica, Lavras, v. 28, n. 3, p. 505-14, 2004.

MARENCO, R.A.; LOPES, N.F. **Plant physiology**: photosynthesis, respiration, water relations and mineral nutrition. Vigosa: UFV, 2007. p.469.

MARCOS FILHO, J. Water/seed relations. In: (Ed.). Seed physiology of cultivated plants. Piracicaba: FEALQ. 2005. p.169-196.

MEDEIROS, A.C.S.; EIRA, M.T.S. Physiological Behavior, Drying and Storage of Native Forest Seeds. Colombo-PR: Embrapa Florestas, 2006. Technical circular, No. 127.

NOBRE, D.A.C., DAVID, A.M.S.S., SOUZA, V.N.R., OLIVEIRA, D., GOMES, A.A.M.,AGUIAR, P.M., MOTA, W.F. Influence of storage environment on the physiological quality of amaranth seeds. **Comunicata Scientiae**, v. 4, n.2, p. 216-219, 2013.

PESKE, S.T.; BARROS, A.C.S.A. **Seed production**. Distance learning specialization course. Brasilia-D.F.: ABEAS. 1998, 76 p.

POPINIGIS, F. **Fisiologia da semente**. 2.ed. Brasilia: AGIPLAN, 1985. 289p.
SILVA, J. S. **Drying and storing agricultural products**. Vigosa: Aprenda Facil, 2008. 560p.

SMANIOTTO, T. A. S.; RESENDE, O.; MARQAL, K. A. F.; OLIVEIRA, D. E. C.; SIMON, G. A. Physiological quality of soybean seeds stored under different conditions. Revista Brasileira de Engenharia Agricola e Ambiental, v.18, n. 4, p. 446-456, 2014.

TOLEDO, F.F.; MARCOS FILHO, J. **Manual de sementes**: tecnologia da produção. Sao Paulo: Ceres, 1977. 233p.

3 Seed conservation

Leticia Winke Dias

Andrea Bicca Noguez Martins

Short-term storage is used to preserve seed quality in the period between harvest and sowing (HARRINGTON, 1972). In the medium term, storage aims to guarantee the annual demand for seeds, making it possible to stock up for years of low production (CARNEIRO & AGUIAR, 1993). In both cases, orthodox seeds can be packed in permeable packaging and stored in a dry chamber or packed in impermeable packaging and stored in a cold chamber. These two storage conditions can be carried out in environments with temperatures above zero (CARNEIRO & AGUIAR, 1993).

In order to preserve seeds in the long term, it is necessary to keep the respiration activity of the seeds at low levels by reducing the ambient temperature and the degree of humidity of the seeds. Most cultivated plants produce seeds whose viability period can be extended by reducing the ambient temperature and the degree of humidity during storage (FREITAS, 2009). These seeds can be stored at sub-zero temperatures. There are limits to storage at sub-zero temperatures only in exceptional cases, for example, when the degree of humidity rises above 15%, which can cause freezing damage to the seeds (GOEDERT, 1980).

3.1.   Storage function

Seed storage begins when they reach physiological maturity, and the biggest challenge is that the seeds still have high physiological quality after a certain period. Therefore, the function of storage is to maintain the quality of the seeds during the period in which they are stored, since their improvement is not possible even under ideal conditions (VILLELA & PERES, 2004).

As seed deterioration is an irreversible process, it cannot be prevented, but it is possible to slow it down by controlling environmental conditions during efficient seed storage (BAUDET, 2003).

It can be said that the function of storage, both for grains intended for food (human or animal) and for seeds, is based on preserving the initial conditions or delaying deterioration by controlling three main factors: seed moisture, relative air humidity and the temperature of the storage environment (MINOR & PASCHAL, 1982; DHINGRA, 1985).

3.2.   Seed moisture content and storage temperature

Seed longevity varies according to genotype, but the period during which physiological potential is preserved depends largely on the degree of humidity and the conditions of the storage environment, with the degree of seed humidity being the characteristic most closely related to the speed of deterioration (MARCOS FILHO, 2005).

During storage, seeds continue with their biological activities, such as respiration, the emission of heat, water vapor and carbon dioxide, the intensity of which depends greatly on the degree of humidity of the seed and the temperature of the environment (CARVALHO & NAKAGAWA, 2000).

The high level of seed moisture during storage is one of the main causes of the loss of germination (DESAI et al., 1997), leading to an increase in the respiration rate and the activity of microorganisms (HARRINGTON, 1972).

The harmful consequences of an increase in the respiration process in a batch of seeds are characterized by wetting and a rise in temperature, which are aggravated when the respiration of microorganisms is taken into account, as well as the respiration of insects that may be associated with the seeds, which can lead to a rapid decline in seed germination and vigour. The increase in the respiration process of seeds also implies an increase in the consumption of reserves, with the consequent loss of dry mass and seed vigor (BAUDET & VILELLA, 2006).

Relative humidity and temperature are the main factors influencing the maintenance of the physiological quality of the seed, particularly its vigor, during storage. The relative humidity of the air is related to the degree of seed moisture, as well as controlling the occurrence of the different metabolic processes it can undergo, while the temperature influences the speed of biochemical processes and indirectly interferes with the degree of seed moisture.

Like humidity, temperature is a very important factor in seed storage. High temperatures accelerate respiration, as well as the activity of microorganisms and insects. In addition, temperature and relative humidity are related, and the effect of one depends on the other (PARRELLA, 2011).

Therefore, temperature can compensate for the effects of high humidity in stored seeds, and seeds with inadequate humidity can be stored for conservation and quality can be maintained through temperature management (BURREL, 1970). The same author showed that barley seeds with high humidity kept for a year at 5 °C.

Studies of the effect of ambient temperature on seed mass have provided important data for improving storage techniques. A cold seed mass is less likely to suffer deterioration. Even if its humidity value is high, low temperatures compensate for the effects of the degree of humidity, both in terms of metabolic processes and attacks by microorganisms, insects and mites (BIZZETTO & HOMECHIN, 1997). Thus, the best conditions for maintaining seed quality are low relative humidity and low temperature, because they keep the embryo at its lowest metabolic activity (CORLETT, 2004).

LITERATURE CITED

BAUDET, L.; VILLELA, F. A. Seed storage. In: PESKE, S. LUCCA FILHO, O. A.; BARROS, A. C. S. A. **Seeds**: scientific and technological foundations. Pelotas: Federal University of Pelotas, 2006. p.428-472.

BAUDET, L. Seed Storage. In: PESKE, S. T.; ROSENTHAL, M. D.; ROTA, G. M. (Ed.) **Sementes:** fundamentos cientificos e tecnologicos. Pelotas: Grafica Universitaria-UFPel, 2003, p.369-418.

BIZZETTO, A., HOMECHIN, M. Effect of storage period and temperature on the physiological and sanitary quality of soybean seeds with high levels of *Phomopsis sojae*. Brasilia, **Revista Brasileira de Sementes**, v. 19, n. 2, p. 295-302, May/Aug. 1997.

BURREL, N. J. The Chiled Storage of Grain Home. **Cereals Authority - Journal Ceres**, v.5, 1970.

CARNEIRO, J.G.A.; AGUIAR, I.B. Seed storage. In: AGUIAR, I.B.; PINA-

RODRIGUES, F.C.M.; FIGLIOLIA, M.B. **Tropical forest seeds.** Brasilia: ABRATES, 1993. p.333-350.

CARVALHO, N. M.; NAKAGAWA, J. **Seeds**: science, technology and production. 4. ed. Jaboticabal: FUNEP, 2000. 588p.

CORLETT, F. M. F. **Physiological quality of annatto seeds *(Bixa orellana* L.) stored in different environments and packaging.** 2004. 94f. Thesis (Doctorate in Seed Science and Technology) - Faculdade de Agronomia Eliseu Maciel, Universidade Federal de Pelotas, Pelotas, 2004.

DESAI, B. B.; KOTECHA, P. M.; SALUNKE, D. K. **Seeds handbook**: biology, production, processing and storage. New York: Marcel Dekker, 1997. 627p.

DHINGRA, O.D. Damage caused by microorganisms during seed storage. **Revista Brasileira de Sementes**, v.7, n.1, p.139-145, 1985.

FREITAS, A. R. Deterioration and storage of vegetable seeds. In: NASCIMENTO, M. W. (Ed.). **Vegetable Seed Technology.** Brasilia: Embrapa Hortaligas, 2009. p.155-182.

GOEDERT, C.O. Germplasm conservation: types of seeds for long-term storage. In: SIMPOSIO DE RECURSOS GENETICOS VEGETAIS, 1980, Brasilia, DF. **Proceedings...** Brasilia, DF: EMBRAPA-CENARGEN: EMBRAPA - DID, 1980. p. 30-32.

HARRINGTON, J.F. Seed storage and longevity. In: KOZLOWSKI, T.T. **Seed biology**. v.3. New York: Academic Press, 1972. p. 145-245.

MARCOS FILHO, J. **Physiology of seeds of cultivated plants**. Piracicaba: Fealq, 2005. 495 p.

MINOR, H.C.; PASCHAL, E.H. Variation in storability of soybeans under stimulated tropical conditions. **Seed Science and Technology,** v.10, p.131139, 1982.

PARRELLA, D.L.N.N. **Seed storage.** Science and Technology Week project for students from the municipalities of Prudente de Morais and Sete Lagoas in the state of Minas Gerais. FAPEMIG, 2011. 16p.

VILLELA, A. F.; PERES B. W. Collection, processing and storage. In: FERREIRA, G. A.; BORGHETTI, F. (Ed.). **Germination:** from basic to applied. Porto Alegre: Artmed, 2004. p.265-281.

4 <u>Seed quality</u>

Aline Klug Radke

Andrea Bicca Noguez Martins

Evaluating the physiological potential of seeds is the main component of a quality control program, as it provides information that identifies and solves problems during the production process, as well as estimating the performance of seeds in the field, ensuring a satisfactory level of performance (AVILA et al., 2006; MARTINS et al., 2014).

When we talk about tests for assessing the physiological quality of seeds for marketing and sowing purposes, we can focus on the germination test, which is carried out under ideal and artificial conditions, allowing the expression of maximum seed quality. However, this test has its limitations, especially when it comes to differentiating batches and the relative delay in obtaining results. Over the years, this has stimulated the development of vigor tests that are reliable and quick, speeding up decisions (BERTOLIN et al., 2011) and complementing the information provided by the germination test.

Therefore, the identification of vigor tests that provide a safe margin as to the behavior of seeds in the field has been a tireless search and a necessity, since the adverse conditions of the environment impose unevenness between the germination test and the field results, thus establishing the need to identify tests that provide conditions equivalent to germination in the field, combined with all the adversities that can affect the performance of a cultivar (MARTINS et al., 2014).

For this reason, over the last few decades, interest in developing appropriate techniques to obtain better information about crops has been a central topic of research (DELL'AQUILA, 2009). The use of seeds with high physiological potential is an important aspect that must be considered in order to increase productivity and, for this reason, seed quality control is becoming increasingly efficient, including tests that quickly assess this aspect, allowing for precise differentiation between batches of seeds that have similar germination (FESSEL et al., 2010).

The period of preservation of the physiological potential depends largely on the degree of humidity and the conditions of the storage environment. The degree of humidity of the seeds is the characteristic most closely related to the speed of deterioration, although the longevity of the seeds varies with the genotype (MARCOS FILHO, 2005).

The harmful consequences of an increase in the respiration process in a batch of seeds are characterized by wetting and a rise in temperature, which are aggravated when the respiration of microorganisms is taken into account, as well as the respiration of insects that may be associated with the seeds, which can lead to a rapid decline in seed germination and vigour. The increase in the respiration process of seeds also implies an increase in the consumption of reserves, with the consequent loss of dry mass and seed vigor (BAUDET; VILELLA, 2006).

4.1. Evaluation of the physiological quality of seeds

Evaluating the quality of seeds after they have been harvested can be done through laboratory tests, which are able to provide reliable data on their ability to produce an adequate and uniform stand of plants in the field, which is the main interest of farmers in the crop establishment phase (PERETTI, 1994).

Determining the germination percentage of a batch of seeds is of significant importance because, through the germination test, the producer will be able to know the maximum percentage of germinable seeds in a batch, guiding the calculation of sowing density. One of the great advantages of the germination test is that it is highly standardized and easily reproducible between laboratories, which makes its results highly reliable (BRASIL, 2009).

The official germination test, which is conducted in the laboratory under favorable conditions for the crop, generally overestimates the physiological potential of seed lots (Radke et al., 2014). However, when conditions are far from ideal, in the field or after a period of storage, batches with similar germination percentages can perform differently due to differences in vigor (LIMA and MARCOS-FILHO, 2011). As such, there is an increasing need to improve the tests used to assess seed vigor, especially with regard to obtaining consistent information and, preferably, in a relatively short period of time (PEREIRA et al. 2011).

Evaluating seed quality using the germination test allows seeds to express their viability under favorable conditions. However, in natural situations, seeds are subjected to a series of adverse conditions, such as variations in soil humidity, radiation and competition, unfavorable conditions that do not always allow them to express their full germination potential (HILHORST et al., 2001). In this sense, the first vigor tests emerged with the aim of identifying batches with the best behavior in the field (PINA-RODRIGUES and VIEIRA, 2004).

The need to obtain reliable results in a relatively short period of time to assess the physiological quality of seeds has increased interest in the use of more sensitive vigor tests to characterize lots, complementing the information from the germination test (CALHEIROS, 2010).

Seed vigor is defined by AOSA (ASSOCIATION OF OFFICIAL SEED ANALYSIS, 1983) as one of the properties of seeds that determines their potential for rapid and uniform emergence with the development of normal seedlings over a wide range of environmental conditions.

Tunes et al. (2012) point out that seed vigor can be defined as the sum of attributes that give the seed the potential to germinate, emerge and quickly result in normal seedlings under a wide range of environmental conditions.

Among the tests used, the accelerated ageing test was initially developed to estimate the longevity of stored seeds, with the aim of standardizing it (Rodo et al., 2000).

The accelerated ageing test is recognized as one of the most widespread tests for assessing the vigour of the seeds of various cultivated species and is capable of providing information with a high degree of consistency (HAMPTON and TEKRONY, 1995; Marcos Filho, 1999). The purpose of this test is to assess the response of seeds to germination after they have been subjected to high temperatures and relative humidity close to 100% for a certain period of exposure.

In species with small seeds, such as vegetables, there have been inconsistent results due to the marked variation in the degree of humidity of the seeds in the samples evaluated after the ageing period. Therefore, alternatives to the accelerated ageing test have been studied, because according to the solution used, specific humidities can be obtained, leading to a reduction in the intensity and rate of water absorption by the seeds, resulting in less deterioration and less variation between the results (JIANHUA and MCDONALD, 1996; TUNES et al, 2012).

LITERATURE CITED:

AVILA, P.F.V.; VILLELA, F.A. & AVILA, M.S.V. Accelerated ageing test to evaluate the physiological potential of radish seeds.

Revista Brasileira de Sementes, vol. 28, n° 3, p.52-58, 2006.

BAUDET, L; VILLELA, F.A. Seed Storage. In: PESKE, S.T.; LUCCA FILHO. O.A.; BARROS, A.C.S.A. (Ed.). **Seeds**: scientific and technological foundations. 2. ed. Pelotas: Ed. Universitaria, 2006, p.427-472.

BERTOLIN D.C., SA M.E., MOREIRA E.R. Parameters of the accelerated aging test for determining the vigor of bean seeds.

Revista Brasileira de Sementes, v. 33, n.1, p.104-112, 2011.

BRAZIL. Ministry of Agriculture, Livestock and Supply. **Rules for Seed Analysis**. Secretariat for Agricultural Defense. Brasilia: MAPA/ACS, 2009. 399p.

CALHEIROS, Veronica Schinagl. **Vigor tests to evaluate the physiological potential of pumpkin seeds *(Cucurbita moschata* Duch.)** - Pelotas, 2010. 34f. ; il..- Dissertation (Master's Degree) - Postgraduate Program in Seed Science and Technology. Eliseu Maciel School of Agronomy. Federal University of Pelotas. Pelotas, 2010.

DELL'AQUILA, A. Development of novel techniques in conditioning, testing and sorting seed physiological quality. **Seed Science and Technology**, v. 37, n.3, p. 608-624, 2009.

FESSEL, S.A., PANOBIANCO M., SOUZA, C.R., VIEIRA, R.D. Electrical conductivity test on soybean seeds stored under different temperatures. **Bragantia**, v. 69, n.1, p.207-214, 2010.

JIANHUA, Z.; McDONALD, M.B. The saturated salt accelerated aging test for small seed crops. **Seed Science and Techology**, v. 25, n. 1, p. 123-131, 1996.

HAMPTON, J. G.; TEKRONY, D. M. **Handbook of vigour test methods**. 3. ed. Zurich: ISTA, 1995.117p.

HILHORST, H.W.M.; BEWLEY, J. D.; CASTRO, R.D.; SILVA, E.A.A.; THEREZINHA, M.; BRANDAO JR., D.; GUIMARAES, R.M.; MACHADO, J.C.; ROSA, S.D.V.F.; BRADFORD, K.J. **Advanced course in seed physiology and technology.** Lavras: UFLA, 2001. p.74.

LIMA, L.B.; MARCOS-FILHO, J. Procedures for conducting vigor tests based on tolerance to heat stress in cucumber seeds. **Revista Brasileira de Sementes [online]**, v. 33, n. 1, p. 45-53. 2011. ISSN 0101-3122. Disponivelem:<http://dx.doi.org/10.1590/S0101-31222011000100005>. Accessed on: May 20, 2014.

MARCOS FILHO, J. Water/seed relations. In: (Ed.). Seed physiology of cultivated plants. Piracicaba: FEALQ. 2005. p.169-196.

MARTINS, A.B.N; MARINI, P.; BANDEIRA, J.M.; VILLELA, F.A; MORAES, D.M. Review: Analysis of seed quality: a nonstop involving activity. **African Journal of Agricultural Research**, v.8, p.114-118, 2014.

PEREIRA, M.F.S; TORRES, S.B.; LINHARES, P.C.F.; PAIVA, A.C.C.; PAZ, A.E.S.; DANTAS, A.H. Qualidade *fisiologica* de sementes de coentro *[Coriandrum sativum* (L.)]. **Revista Brasileira Plantas Medicinais**, v.13,

Special, p.518-522, 2011.

PERETTI, A. **Manual para analise de semillas.** 1ed. Buenos Aires: Hemisferio Sur, 1994. 282p.

PINA-RODRIGUES, F.C.M.; VIEIRA, J.D. Germination test. In: PINA- RODRIGUES,

F.C.M. **Manual de analise de sementes florestais**. Campinas: Fundapao Cargill, 1988. p. 70-90

RADKE, A.K.; REIS, B.B.; ALMEIDA, A.S.; MENEGHELLO, G.E.; TUNES, L.M.; VILLELA, F.A. Alternative methodologies to test seed vigor in lettuce. **Enciclopedia biosfera**, v.10, n.19, p. 94-101,2014.

RODO, A.B.; PANOBIANCO, M.; MARCOS FILHO, J. Alternative methodology of the accelerated aging test for carrot seeds. **Scientia Agricola**, Piracicaba, v.57, n.2, p. 289-292, 2000.

TUNES, L. M., TAVARES, L. C., DE ARAUJO RUFINO, C., BARROS, A. C. S. A., MUNIZ, M. F. B., & DUARTE, V. B. Accelerated aging of broccoli seeds *(Brassica oleracea* L. var. italica Plenk)= Accelerated aging of broccoli seeds *(Brassica oleracea* L. var. italica Plenk). **Bioscience Journal**, *v.28, n.2,* p. 173-179, 2012.

5 <u>Seed deterioration</u>

Fernanda da Motta Xavier

Bruna Bezerra

Andrea Bicca Noguez Martins

Seeds reach their maximum quality at the point of physiological maturity, after which they are subject to a series of degenerative changes of biochemical, physiological and physical origin. These changes characterize the deterioration process, which is associated with a reduction in vigor and loss of germination capacity (FREITAS, 2009).

The term deterioration refers to any and all physiological, biochemical, physical or cytological alterations, starting at physiological maturity, at a progressive rate, determining a drop in quality and culminating in the death of the seed (MARCOS FILHO, 2005).

Each and every seed in storage suffers deterioration, which can be faster or slower, depending on the environmental characteristics and the characteristics of the seed itself. Generally, reducing the light, temperature and humidity of both the seed and the environment reduces their metabolism and the microorganisms that deteriorate them are out of action, increasing their longevity (VIEIRA et al., 2001).

The period that a seed can live is determined by its genetic characteristics and is called longevity, while the period that the seed actually lives is determined by the interaction between genetic and environmental factors. This period is called viability. It is practically impossible to accurately determine the true longevity period of the seeds of a species, however, the length of time that a seed actually lives within its longevity period depends on factors such as the genetic characteristics of the parent plant; the vigor of the parent plants and the climatic conditions during maturation (CARVALHO & NAKAGAWA, 2000).

The manifestation of seed deterioration is closely associated with storage. However, theoretically this process begins at physiological maturity and can be accelerated at any stage after maturity, possibly extending to the post-sowing period. It cannot be denied, however, that it is detected more frequently during storage. The loss of germination power is the final consequence of seed deterioration, i.e. considerable deterioration occurs before there is a decrease in germination percentage (MARCOS FILHO, 2005).

The main alterations involved in seed deterioration include the depletion of food reserves, alterations to the chemical composition, such as the oxidation of lipids and the partial breakdown of proteins, alterations to cell membranes, with a reduction in their integrity, enzymatic alterations and alterations to nucleotides (FERREIRA; BORGHETTI, 2004).

5.1.   Causes of deterioration

Seeds deteriorate under all environmental conditions, whether in common stores or germplasm banks. This process, also known as seed ageing, is of great importance because its effects are often only felt in the field, after the seeds have shown poor performance. The reduction in the physiological potential of the seeds can be seen in the reduction in germination, speed and uniformity of seedling emergence.

Other events observed in seed deterioration include changes in genetic material, low enzyme activity and lower ATP production. There are also several references to a reduction in DNA synthesis and respiratory activity, as well as the sensitivity of mitochondria to deterioration. In addition to these biochemical manifestations, changes in the structure of the cell membrane system are highlighted. It is known that organelles in different parts of cell tissues have specific functions, differing in shape, metabolism, function and sensitivity. Macromolecules, which are essential for germination, are degraded during ageing (MARCOS FILHO, 2005). According to Marcos Filho (2005), some of these key proteins can be indicators of seed quality (vigor), even when there is a decrease in seed reserves.

LITERATURE CITED

CARVALHO, N.M.; NAKAGAWA, J. **Seeds**: science, technology and production. 4. ed. Campinas: Cargill Foundation, 2000. 588p.

FERREIRA, A. G.; BORGHETTI, F. **Germination**: from basic to applied. Porto Alegre: Artmed, 2004. 323 p.

FREITAS, A. R. Deterioration and storage of vegetable seeds. In: NASCIMENTO, M. W. (Ed.). **Vegetable Seed Technology.** Brasilia: Embrapa Hortaligas, 2009. p.155-182.

MARCOS FILHO, J. Water/seed relations. In: (Ed.). Seed physiology of cultivated plants. Piracicaba: FEALQ. 2005. p.169-196.

VIEIRA, A H.; MARTINS, E.P.; PEQUENO, P. L de L.; LOCATELLI, M.; SOUZA, M.G. de. **Forest seed production techniques.** Porto Velho: Embrapa Rondonia, 2001.4p. (Technical Circular, 205).

6 Longevity of Seeds

Andre Pich Brunes

Andrea Bicca Noguez Martins

After reaching physiological maturity, all seeds gradually lose their vitality (in the process of deterioration), depending on the species considered, the chemical composition of the seed, and the conditions under which they were produced and stored. Longevity corresponds to the maximum period in which seeds remain alive when stored under ideal environmental conditions, and species show natural variability (MARCOS FILHO, 2005). According to their longevity, seeds can be classified as short, medium or long-lived. In general, long- and medium-lived seeds have an increasing longevity as the ambient temperature and water content decrease.

The longevity of orthodox seeds (those that can be dried) can be significantly extended by drying them at water contents of up to 5% to 8% and by packaging them in waterproof containers. Short-lived seeds, on the other hand, often have a very short lifespan of weeks or months. Also known as recalcitrant, they do not tolerate desiccation and/or are susceptible to low temperatures.

The storage potential of orthodox seeds is influenced by a number of factors, apart from longevity, including initial quality, which is determined by the environmental conditions prevailing during the seed filling phase and the period between physiological maturity and harvest, mechanical damage during harvest and processing, especially on conveyors, and thermal damage during drying.

Mechanical damage is a serious problem that can reduce seed quality. As a result, there is a reduction in vigor, reduced germination, increased infection and proliferation of pathogens and lower storage potential. Broken seeds can be easily removed during processing. However, seeds with small cracks or dents remain in the batch, becoming sources of deterioration.

LONGEVITY	SPECIES
Long (over 15 years)	Alfalfa, cotton, oats, carrots, barley, beans, flax, melons, corn, sorghum, tomatoes, wheat.
Medium (between 3 and 15 years)	Rice, ryegrass, string beans, soybeans.
Short (less than 3 years)	Cocoa, rubber, carambola, coffee and mango trees

Source: Seed News Magazine, July/Aug. 2009

6.1. Seed storage problems

Storage conditions are one of the main factors in guaranteeing seed quality. Often the farmer goes to all the trouble of selecting the best seeds, but due to a lack of care in storage, all the effort is lost. Poor storage causes the following problems.

6.1.1. Mildew

If seeds are not dried to the correct moisture level before being sealed in suitable containers, they can rot, which happens quite often. A simple test:

after drying and placing in closed glass jars or plastic bags, the appearance of condensation inside the containers within a few hours indicates the need for further drying. The use of silica gel is necessary.

6.1.2.　　Insect attack

Insects that may have escaped attention can cause damage to stored seeds. Alternative methods such as the use of inert materials to control these pests have been studied by researchers. One example would be the use of diatomaceous earth. It is a light material with a low apparent specific mass, whose color varies from white to dark grey. Its main component is silica, which is found in hydrated form, as well as aluminum, iron, magnesium and sodium and, because it is practically non-toxic, it can be easily handled by farm and storage workers, providing protection for the seeds. Using a few pinches of diatomaceous earth is a safe, cheap and non-toxic way to protect seeds from insect damage. Be sure to lightly coat all the seeds before final sealing and storage. Diatomaceous earth is available in most garden stores.

6.1.3.   Rodent attacks

Seeds not stored in glass or metal containers can provide a real feast for mice and small vermin. Make sure to keep all seeds in metal or glass containers and properly labeled.

Tips

✓ Storage does not improve the qualities of the seed lot, it only maintains them.

✓ Temperature and humidity are the two most important factors in storage.

✓ The greater these factors, the greater the physiological activity of the seed.

✓ The best conditions for storing orthodox seeds are cold and dry climates.

✓ Immature and damaged seeds are not suitable for storage, while mature and undamaged seeds are ideal.

✓ Storage potential varies according to the species.

Printed by Books on Demand GmbH, Norderstedt / Germany